Beatriz Dantas Guimarães

Molecular Biology Techniques: Polymerase Chain Reaction (PCR)

Beatriz Dantas Guimarães

Molecular Biology Techniques: Polymerase Chain Reaction (PCR)

Amplification of the dengue virus envelope protein gene using the Polymerase Chain Reaction technique

ScienciaScripts

Imprint

Cover image: www.ingimage.com

This book is a translation from the original published under ISBN 978-3-330-76526-9.

Publisher:
Sciencia Scripts
is a trademark of
Dodo Books Indian Ocean Ltd. and OmniScriptum S.R.L publishing group

120 High Road, East Finchley, London, N2 9ED, United Kingdom
Str. Armeneasca 28/1, office 1, Chisinau MD-2012, Republic of Moldova, Europe
Managing Directors: Ieva Konstantinova, Victoria Ursu
info@omniscriptum.com

Printed at: see last page
ISBN: 978-620-8-51549-2

CONTENTS

CHAPTER 1

INTRODUCTION

Polymerase chain reaction technology was conceived in the 1980s and its main application is the possibility of amplifying a certain segment of DNA through enzymatic synthesis in vitro, producing millions of copies of the specific segment selected (CARUSO, 2007; ZAHA et al., 1996). PCR involves three stages, which make up a basic PCR replication cycle: denaturation of the DNA template, hybridisation with primers that delimit the region of interest to be duplicated, extension of the selected segment using the DNA-polymerase enzyme.

In addition to the process of amplifying specific DNA sequences, several other applications can be attributed to and aided by this technique: the amplification of a gene using PCR is a key step in obtaining the amount of DNA needed to be applied in the molecular cloning stage, and real-time PCR is now used to quantify gene expression (CARUSO, 2007).

Gene cloning aims to introduce a selected gene into bacteria that are identical to each other (NASCIMENTO et al., 1999) and consists of isolating and propagating identical DNA molecules, being the central technique for establishing the recombinant DNA technology method (CARUSO, 2007). The technique requires the presence of a cloning vector capable of amplifying a fragment of DNA cloned into it - such as the genes obtained by PCR - the plasmid being one of the most relevant cloning vectors. The amplified gene must be attached to the vector and the hybrid molecule must be introduced into a host cell, usually bacterial cells (LIMA, 2008). The manipulation of specific genes using bacterial transformation containing the transgene DNA molecules allows, at the end of the steps, the expression and obtaining of hybrid proteins.

Dengue is an acute infectious disease that is a major public health problem worldwide (GUBLER, 1998) and its etiological agent, the dengue virus (DENV), is classified as an arbovirus belonging to the Flaviviridae family, Flavivirus genus. They are single-stranded RNA viruses, comprising four viral serotypes of epidemiological importance (DENV-1, DENV-2, DENV-3 and DENV-4), genetically related, sharing approximately 65% of the nucleotide sequence of their genomes (MONDINI et al., 2009; WHITEHEAD et al., 2007; MURREL et al., 2011). The DENV genome encodes a single polyprotein, which is subsequently cleaved into seven non-structural proteins (NS1, NS2A, NS2B, NS3, NS4A, NS4B, NS5) and three structural proteins that make up the virus (capsid protein C, membrane protein M and envelope protein E) (CHAMBERS et al., 1990).

Although they are very close genetically, the DENV serotypes are antigenically distinct, mainly due to the differences found in the amino acid sequences present in the E protein of the four viral serotypes (SHRESTHA et al., 2010).

The four dengue serotypes can cause anything from an undifferentiated febrile illness to more severe forms such as dengue haemorrhagic fever, all of which confer permanent immunity to the serotype responsible for the infection. However, infection does not confer cross-immunity against the other serotypes (PANAMERICAN HEALTH ORGANISATION, 1995),

Since the most severe forms are mainly associated with the immune response in cases of reinfection by a different serotype

(MONGKOLSAPAYA et al., 2003), distinguishing the infecting serotype is important, but distinguishing by PCR or viral isolation combined with immunological methods for identifying strains using antibodies to the non-structural NS1 protein is time-consuming, difficult and has low sensitivity (ARAÚJO, 2015; BRASIL, 2013; CARDOSA, 1998).

The E gene, the main target of this study, is the most widely used for molecular studies of DENV, due to the importance of the E glycoprotein encoded by this gene (CHAMBERS et al., 1990; LEITMEYER et al., 1999). The E protein is the largest protein on the surface of the viral particle, made up of three distinct domains, with important functions in viral biological activity, such as interaction with host cell receptors and entry of the virus into the cell. Studies suggest that the E protein is the most suitable for choice in viral serotype diagnosis (BARROS, 2007) because it is exposed on the surface of the virus, is responsible for interaction with cell receptors and represents the main target of neutralising antibodies (JAISWAL et al., 2003), emphasising the importance of this research.

This work is part of a project whose final objective is to express the E protein of the 4 different viral serotypes, using the recombinant molecule of the E gene inserted into the pGEX-2T cloning vector, by means of gene cloning, with the aim of developing an ELISA immunodiagnostic test to distinguish the serotypes with monoclonal antibodies produced from the E protein, in cooperation with the department of Electrical Engineering at the Federal University of Campina Grande (UFCG).

This work was carried out on the premises of the Três Marias Laboratory Centre at the Centre for Biological and Health Sciences (CCBS) of the State University of Paraíba (UEPB), with the aim of establishing techniques for amplifying the E gene of DENV-2 and DENV-4: these steps are essential for cloning and expressing the E protein of these serotypes. The subsequent steps for the molecular cloning technique are underway at the same location.

CHAPTER 2

LITERATURE REVIEW

2.1 Polymerase chain reaction

The polymerase chain reaction (PCR) is a procedure that makes it possible to amplify specific DNA fragments, starting from a target DNA, and is a widely used technique for directly cloning DNA with genes of interest. PCR is a very sensitive technique, allowing the detection and amplification of a single DNA molecule in a sample (ALBERTS et al., 2009).

To apply the technique, you first need the double strand of DNA - which will serve as a template for amplification - and a pair of specific oligonucleotides (primers), which will initiate DNA synthesis. The primers are single strand segments that are complementary to the template strands in positions that define the ends of the segment to be amplified. PCR also requires three other components: DNA polymerase, dNTPs, buffer and salts (MgCl) (LIMA, 2008; NELSON et al., 2014; ALBERTS et al., 2009).

In the presence of a thermostable DNA polymerase, DNA precursors (the four deoxyribonucleoside phosphates - dNTPs: dATP, dCTP, dGTP, dTTP), ideal pH conditions (using buffer), salinity and the enzyme cofactor (bivalent magnesium ions - Mg+2), the synthesis of the new strands begins. The volume of the reaction varies depending on the objective, and is carried out in volumes of 10 to 50p,l (LUDWIG, 2010; UHLMANN et al., 1998).

The PCR method consists of repetitive cycles of 3 steps:

Step 1 - DNA DENATURATION: The two DNA chains are held together in the double helix by relatively weak hydrogen bonds, which break at high temperatures. The mixture of components is heated to denature the DNA (usually at 94°C for 1-2 minutes), separating the double strand and forming two single strands that will be used as templates for the primers and DNA polymerase. The phosphodiester bonds between phosphate and deoxyribose are stronger covalent bonds and remain intact (LUDWIG, 2010; UHLMANN et al., 1998; LIMA, 2008).

Step 2 - PRIMER RINGING: The temperature is then cooled to allow the primers to hybridise (ring) to the complementary sequences in the DNA, so that they can renature with the DNA. The annealing temperature varies between 50 and 60 °C, and depends on the average denaturation temperature of the primers *(melting* temperature - Tm). *Primers* unique to each end of the target sequence are needed to amplify only the sequence of interest, which is why it is the most important variable in determining the specificity of PCR, so the temperatures and annealing times vary according to the sequence to be amplified (LUDWIG, 2010; Kim et al., 2002).

Step 3 - POLYMERISATION: In this step, the mixture is heated again to the optimum replication temperature (approximately 72°C), so that the DNA fragment is selectively replicated in the annealing region of each *primer*, by the addition of dNTPs by a DNA polymerase enzyme. Depending on the size of the segment to be amplified, the extension time will vary from 30s to 5min (LUDWIG, 2010; Kim et al., 2002; LIMA, 2008).

The heating, cooling and replication cycle is repeated for 20 to 30 cycles in thermal cyclers, which automatically control and alternate temperatures during programmed periods of time, for the appropriate number of cycles. Each cycle doubles the amount of DNA synthesised in the previous cycle, and the new fragments generated serve as a template for the next cycle. In this way, several cycles result in the exponential accumulation of the DNA fragment flanked by the two primers (LUDWIG, 2010; NELSON et al., 2014).

As each cycle requires a heat treatment to separate the strands of the double helix from the DNA mould, PCR uses a thermostable DNA polymerase, isolated from a thermophilic bacterium Thermus aquaticus (Taq DNA polymerase), so that it is not denatured by the heating of each cycle (LUDWIG, 2010; NELSON et al., 2014; ALBERTS et al., 2009).

The PCR result (amplicon) is generally analysed by agarose or polyacrylamide gel electrophoresis, and the type of matrix used will depend on the size of the DNA segments to be separated and visualised. Due to the difference in pore size of the matrices, agarose gels are generally used to separate fragments ranging from 0.2 kb to 50 kb (1 kb = 1000pb - base pairs) and polyacrylamide gels for fragments up to 1kb.

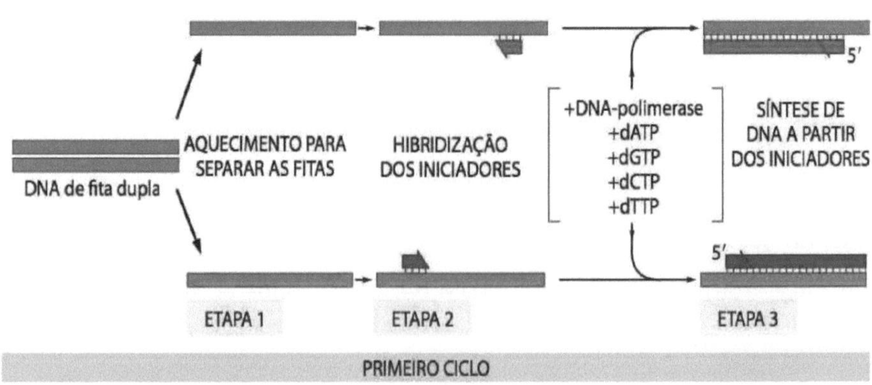

Figure 1 - Stages of the first cycle of DNA amplification using the PCR technique.

Source: ALBERTS, 2009

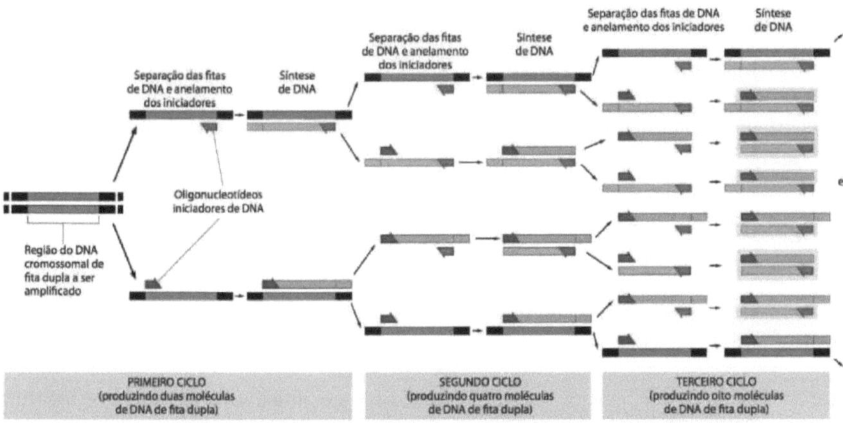

Figure 2 - DNA amplification results after several cycles.

Source: ALBERTS, 2009

In this work, the protolocates for amplifying the E gene of DENV-2 and DENV-4 were established, using specific primers that included this gene encoding the E protein.

2.2 Molecular cloning

Molecular cloning is the central technique of recombinant DNA technology, which consists of isolating and propagating DNA molecules with the recombinant gene of interest in bacterial cells that are genetically identical to the initial one, isolating them in colonies (LIMA, 2008). It basically consists of two stages:

STEP 1 - Linking a DNA fragment with the gene of interest (called an insert) with another DNA molecule, the cloning vector, to form a chimera or recombinant DNA (BROWN, 2003).

STEP 2 - The recombinant molecule is transported into a compatible host cell, usually a bacterium, and the process called transformation takes place. The cell that receives the DNA is called the transformant (BROWN, 2003). Under ideal conditions, a single transformant undergoes numerous cycles of cell division, producing colonies with thousands of copies of the recombinant DNA (LIMA, 2008; CARUSO, 2007).

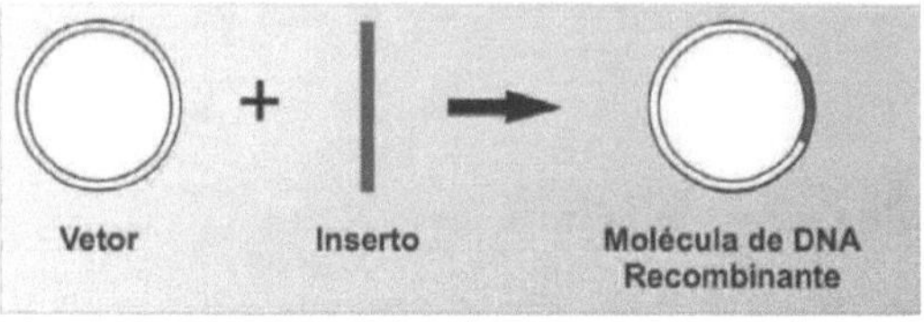

Figure 3 - Ligation of the cloning vector with the insert (amplified gene of interest), producing a recombinant DNA molecule.

Source: LIMA, 2008.

To carry out stage 1 of molecular cloning, two DNA molecules are essential: the insert, containing the gene of interest; and a molecule capable of amplifying this information in hundreds of copies: a cloning vector.

2.2.1 Cloning vectors

The basic types of vectors used in the recombinant DNA methodology have special characteristics that make them excellent cloning vehicles, the most commonly used vectors being bacterial plasmids.

Plasmids are small, circular, double-stranded DNA molecules that exist in bacteria outside the bacterial chromosomes and are capable of replicating independently. Plasmids vary in structure, size, replication mode and ability to propagate in different bacteria. They are used as cloning vehicles as long as they enable the amplification of a gene of interest inserted into them, contain elements necessary for their replication and at least one gene that confers antibiotic resistance (NASCIMENTO, 1999; CARUSO, 2007; LIMA, 2008).

To be considered a good cloning vector, a plasmid must contain three basic characteristics (NASCIMENTO, 1999; BROWN, 2003;

WEAVER, 2001):

1) **REPLICATION ORIGIN** - DNA sequence that allows the vector to replicate independently in the host cell.

2) **CLAIMING SITES** - two or more cleavage sites for restriction endonucleases, into which the DNA fragment is inserted, without interfering with the functionality of the plasmid. The set of these sites is called Multiple Cloning Sites (MCS).

3) **GENE THAT CODES A PRODUCT THAT DISTINCTS THE TRANSFORMED CELL FROM THE NON- TRANSFORMED CELL** - such as genes that confer resistance to antibiotics.

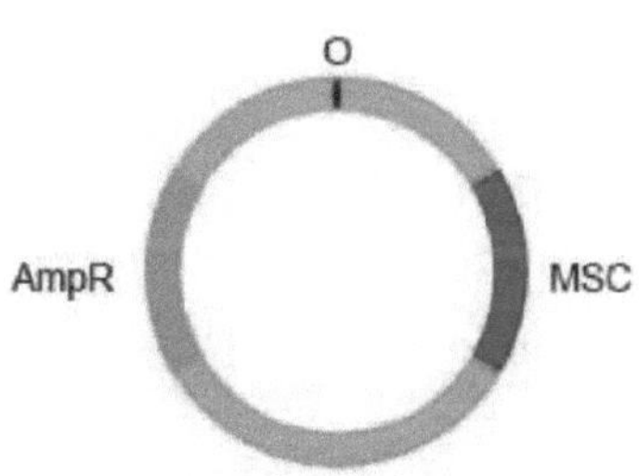

Figure 4 - Molecular structure of a typical plasmid used in molecular cloning. The gene that confers resistance to the antibiotic Ampicillin is represented by AmpR, the region of multiple cloning sites by MSC and the origin of replication by O.

Source: CARUSO, 2007.

The amplified DNA gene, such as the E protein gene, which is the focus of this work, can be inserted into a linearised plasmid and treated with the enzyme alkaline phosphatase - which prevents the vector from rewiring - together with the enzyme T4 DNA ligase, which allows them to be linked, generating a transgenic DNA molecule.

2.3 Structural and genomic organisation of DENV

DENV belongs to the Flaviviridae family, genus Flavivirus, with four antigenically distinct serotypes: DENV-1, DENV-2, DENV-3 and DENV-4 (WESTAWAY et al., 1985). There is also the DENV-5 serotype, but only the first four have epidemiological importance to date (BARROS et al., 2008; MACIEL et al., 2008; MUSTAFA et al., 2015). It is a spherical virus, 40 to 50 nm in diameter, containing an electron-dense core of 30 nm surrounded by a bi-lipid envelope, where the E proteins in dimers (envelope) and the M protein (membrane) are anchored (LINDENBACH et al., 2007; WESTAWAY et al., 1985).

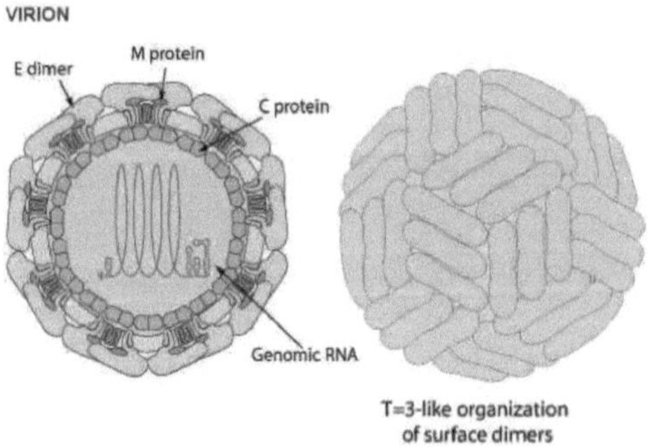

Figure 5 - Structure of the viral particle (DENV). The virus is surrounded by a bilipid envelope in which proteins E and M are anchored, as well as capsid protein C, with affinity for the viral RNA genome.

Source: http://viralzone.expansy.org/viralzone/all_by_species/43.html.

The virus has capsid protein C, which is highly basic and has affinity for the viral genome, which consists of single stranded RNA with positive polarity and a size of approximately 11kb (ALCATRAZ-ESTRADA et al., 2010). This genome has around 11,000 nucleotides and only one open reading frame (ORF) (CHAMBERS et al., 1990), in which it encodes a single polyprotein that is subsequently cleaved by cellular proteases into the 3 proteins

structural proteins (protein C, protein M and protein E) - located in the 5' portion - and in 7 non-structural proteins (NS1, NS2A, NS2B, NS3, NS4A, NS4B, NS5) - in the 3' portion (BARROS, 2007; YAP et al., 2007).

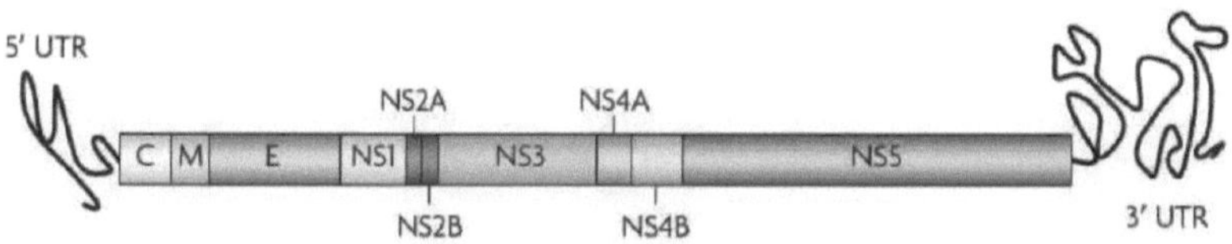

Figure 6 - DENV genomic organisation.
Source: GUZMÁN, 2010.

2.3.1 Protein E

The 50-55 kDa envelope glycoprotein E is the main component of the virus surface, and the main biological properties of DENV are related to this protein. It is responsible for phenotypic, antigenic, virulence and tropism properties, as well as inducing fusion with the membrane and binding to cell receptors (OLIVEIRA, 2013), being essential in the phases of absorption and penetration of the virus into the cell (HENCHAL et al., 1990). According to Bricks (2004), antibodies directed at this protein determine the longest-lasting protective immunity.

The E protein has three structural and functional domains (DI, DII and DIII), which are related to the presence of antigenic regions and specific epitopes (CRILL & CHANG, 2004). Domain I is the central N-terminal region, domain II is the dimerisation region of the protein and is involved in virus-mediated membrane fusion. Domain III is the region responsible for binding the protein to the host cell receptor, and is also the most targeted domain for antibody neutralisation (GUZMÁN; KOURÍ, 2004).

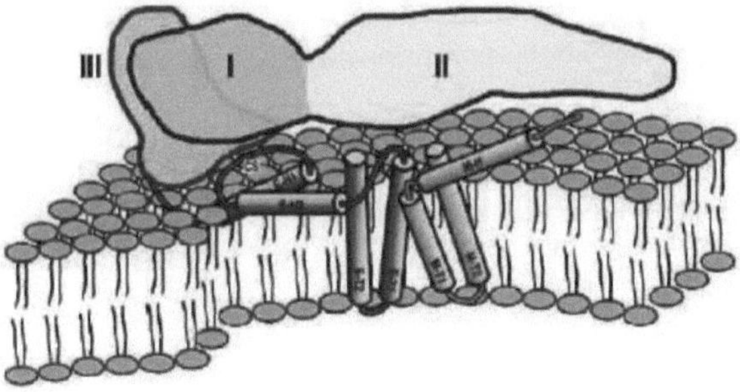

Figure 7 - Representation of the three domains of protein E. Domain I is shown in pink, domain II in yellow and domain III in blue.

Source: ZHANG, 2003.

After glycoprotein E binds to the surface receptors, the viral particle is internalised in an endosomal compartment. Under the low pH conditions found in endosomes, the homodimers of protein E dissociate into monomers, then re-associate into irreversible trimers to expose the domain II fusion peptide. The trimer extends towards the endosome membrane, and the fusion peptide inserts into it. The fusion of the endosomal and viral membranes enables the viral nucleocapsid to penetrate the host cell and the viral genome to be denuded (MUKHOPADHY et al., 2005).

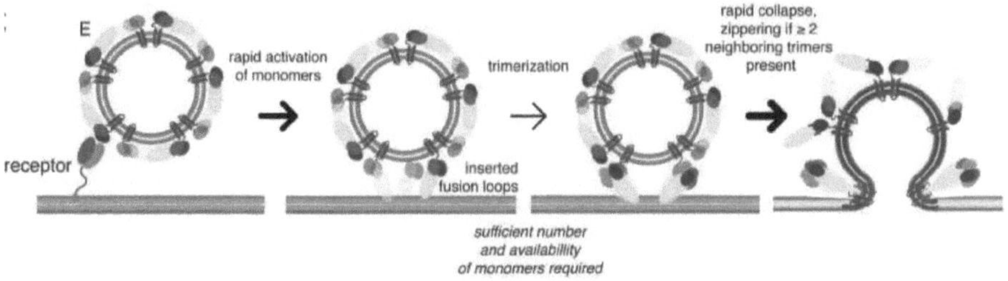

Figure 8 - Virus internalisation mediated by the E protein.
Source: CHAO, 2014.

Most virus neutralising antibodies are against specific epitopes of the E protein (CARDOSA et al., 1998). Monoclonal antibodies to the E protein have been shown to be able to neutralise the dengue virus in vitro (CLYDE et al., 2006) and their relationship with cell receptors involved in fusion with the membrane and entry of the virus into the cell have already been elucidated for different serotypes of the dengue virus, and are the subject of many research studies (ARAÚJO, 2015; KOSTYUCHENKO et al., 2014).

CHAPTER 3

METHODOLOGY

For the cloning process, a recombinant plasmid with the E gene inserted is required. This requires PCR amplification of the E gene from DENV-2 and DENV-4 in order to proceed with the subsequent cloning steps. The methods for establishing the E gene amplification technique are described below.

3.1 Amplification of the pGEX-2T vector

The cloning vector used, pGEX-2T, consists of a circular DNA molecule, as shown in figure 9 below:

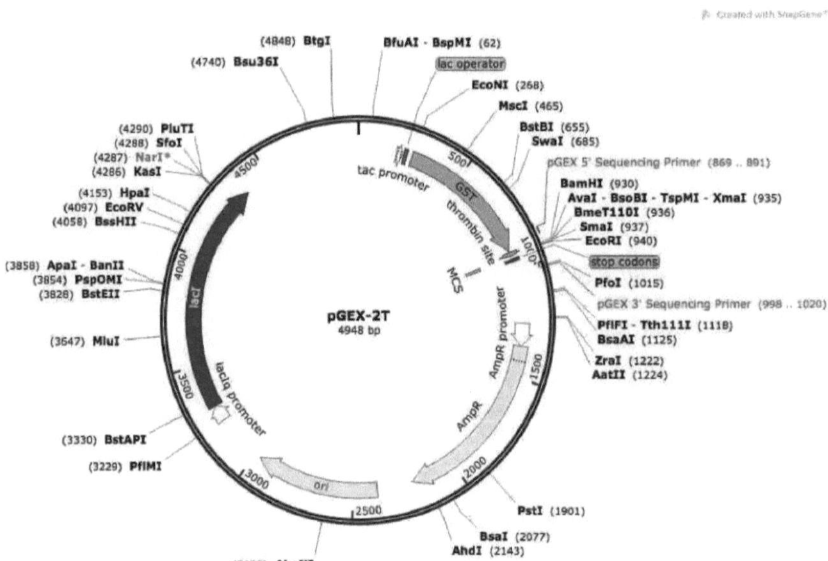

Figure 9 - pGEX-2T cloning vector plasmid. It can be seen that this plasmid contains all the essential characteristics of a vector: origin of replication, cloning site (Sma I and Eco RI), numerous restriction sites (represented by the acronyms in bold) and a gene that provides resistance to ampicillin.

The pGEX-2T vector is a protein expression vector. The Eco RI and Sma I restriction sites allow the circular DNA of the plasmid to be linearised so that the gene of interest can be inserted into the vector and the transgene formed. By inserting the DNA into these sites, it is possible to generate a reading text for a chimeric protein, encoded by the Glutathione-S-Transferase (GST) gene and the DNA of the inserted gene. The GST is used to purify the expressed protein.

The pGEX-2T vector had previously been treated with the restriction enzyme Sma I (937 bp) for later use in cloning. A complementary procedure to the digestion was carried out to prevent empty rewiring of the vector: the removal of phosphorus from the free 5' ends of the linearised vector, by the action of the enzyme alkaline phosphatase. In this way, pGEX-2T, properly linearised and treated, remains suitable for transgene insertion after appropriate amplification.

In this study, the primers selected were aimed at amplifying just one fragment of the circular pGEX-2T vector, measuring 1146 bp, in order to establish a control PCR. The primers used for amplification and their respective denaturation temperatures are shown in the table below:

Table 1 - *Primers* used to amplify the 1146 bp fragment of the pGEX-2T vector and their respective denaturation temperatures.

Forward *primer*	**Reverse *primer***	***Melting* temperature**
pGEX-F2	pGEX-R2	pGEX-F2 (61.4°C) pGEX-R2(61.4°C)

Nucleotide sequences of the primers:

pGEX-F2: TCAGGTGGCACTTTTCGGGG

pGEX-R2: GAGCGTCAGACCCCGTAGAA

The nucleotide sequences of pGEX-2T, as well as the location of the selected primers and the genome fragment amplified with these primers can be found in **APPENDIX A**.

The components for mixing the reaction and their respective quantities, totalling 50 gl, are listed below:

39.5 µl H2O;
5 µl Buffer;
1.5 µl MgCl2 (concentration - 50mM);
1µl dNTPs (concentration - 10mM);
0.5 µl pGEX-2T (template);
1 µl Forward primer (concentration - 20mM);
1 µl Reverse primer (concentration - 20mM);
0.5 µl Taq polymerase.

3.2 Amplification of the E gene: DENV-2 and DENV-4

The PCR (polymerase chain reaction) technique was used to amplify the E gene of the viral envelope protein. The DNA molecule used in this work was obtained from FIOCRUZ in Recife, containing the E protein gene of DENV-2 and DENV-4, with around 1500 base pairs - according to data available on the NCBI platform (references: Dengue virus type 2 isolate CEA2440 polyprotein gene, partial cds GenBank: AY775303.1; Dengue virus 4 strain SPH317947 envelope glycoprotein gene, partial cds. and envelope (E) protein [Dengue virus 4] NP_740317.1).

3.2.1 Primers - DENV-2 and DENV-4

The forward and reverse primers for DENV-2 and DENV-4 were used in the following combinations:

Table 2 - Primers used to amplify DENV-2 and DENV-4 and their respective denaturation temperatures.

Forward *primer*	**Reverse *primer***	**Combination**	***Melting* temperature**
DE2-F3	DE2-R2	DE2-F3R2	DE2-F3 (56.42°C)
DE2-F4	DE2-R3	DE2-F4R3	DE2-F4 (63.73°C)
DE4-F2	DE4-R2	DE4-F2R2	DE4-F2 (59.82°C)
DE4-F3	DE4-R3	DE4-F3R3	DE4-F3 (63.73°C)

Nucleotide sequences of the primers:

DE2-F3: A ATG CGT TGC ATA GGA ATT TCA AAT
DE2-R2: GA GCT ATG GTG CAG GCT TGA
DE2-F4: GTG GGG AGG CAA GAG AAA GGG
DE2-R3: CGC CAT CAA CTT TGG CAA GAG
DE4-F2: A ATG CGA TGC GTA GGA GTA GG
DE4-R2: G GGC TTC ACA GTT GGC GCC
DE4-F3: CCT AAT GAT GCT GGT CGC CCC
DE4-R3: GGA TGC GCC ATC AAC TTT GGC AAG

.2.2 The reaction mixture and the types of PCR carried out

It is important to note that the combination of primers DE2-F4R3 and DE4- F3R3 starts synthesis in the prM gene (membrane glycoprotein precursor gene), externally to the E gene (external primers), compared to the combination of DE2-F3R2 and DE4-F2R2 (internal primers), which start and finish synthesis directly in the E gene.

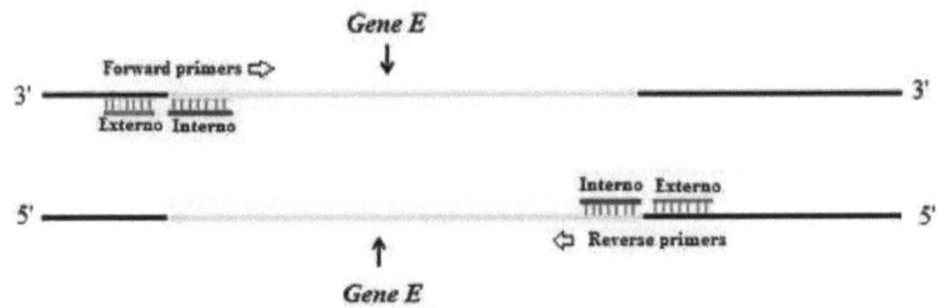

Figure 10 - Schematic of the external *primers* and internal *primers* used to amplify the E gene.

Table 3 below shows all the reaction arrangements made in the PCR of the E gene, with the respective primers and template DNA:

Table 3 - *Deprimer* combinations and PCR reaction arrangements for the E gene.

DNA (template)	*Primers*
DENV-2	DE2-F3R2 internal *meimers*)
DENV-2	DE2-F4R3 external *epsimees*)
PCR product (DE2-F4R3)	DE2-F3R2 internal *mrimers*)
DENV-4	DE4-F2R2 internal *meimers*)
DENV-4	DE4-R3F3 external *epsimees*)
PCR product (DE4-F3R3)	DE4-F2R2 (internal *primers*)

The nucleotide sequences of part of the DENV-2 and DENV-4 genomes, as well as the location of the internal and external primers selected for amplification of the E gene in the genome can be found in **APPENDIX B** and **APPENDIX C**, respectively.

This technique is called Nested PCR. Since larger DNA fragments - resulting from the PCR of the external *primers* - would be used as a template, the E gene would be more likely to be successfully amplified, increasing the sensitivity of the technique.

Furthermore, the use of *Nested PCR* often - but not always - makes it possible to obtain a greater quantity of amplified final product - in this case, the E gene - from both amplification with external *primers* and amplification with internal *primers* using the previous PCR product as a template.

The same quantities for the reaction mixture as described above were used for this procedure, as described below:

39.5 µl H2O;
5 µl Buffer;
1.5 µl MgCl2 (concentration - 50mM);
1 µl dNTPs (concentration - 10mM);
0.5 µl DNA (template);
1 µl Forward primer (concentration - 20mM);
1 µl Reverse primer (concentration - 20mM);
0.5 µl Taq polymerase.

3.3 Thermocycler

3.3.1 Ringing temperature

The temperature used to anneal the primers was established according to the formula for calculating the average denaturation temperature of the primers (melting temperature):

$$\mathbf{TM = 69.3 + GC\% \times 0.41 - (650 : N)}$$

Where,

TM = melting temperature

CG% = percentage of cytosine and guanine in the primer nucleotide sequence

N = total nucleotides in the primer

In order to programme the thermal cycler and carry out all the PCRs in the same programme, the value for the annealing temperature (55°C) was determined so that it was below the lowest average denaturation temperature among the primers, i.e. below 56.42°C (DE2-F3).

3.3.2 Programming the thermal cycler

All the amplifications of the DNAs (DENV-2, DENV-4, pGEX-2T) with their respective primers and components, in the appropriate quantities, took place in the same thermal cycler programming cycle, as shown in table 4:

Table 4 - Programming the temperature cycles in the PCR thermal cycler for amplification of the dengue E gene and the pGEX-2T cloning vector.

	Home	**30 cycles**			**End of story**	
Temperature	94°C	***Denaturation of D NA***	***Ringing of primers***	***Polymerisation***	72°C	4°C
		94°C	55°C	72°C		
Time (seconds)	60s	30s	30s	90s	300s	00

3.4 Electrophoresis

Gel electrophoresis was used to separate and visualise the fragments amplified by the PCR technique. Agarose was selected because it contains larger pores in its matrix and is more suitable for separating larger fragments (200-5000 base pairs), as is the case with the segments amplified in this study (approximately 1500 base pairs of the E gene; 1146 base pairs of the pGEX-2T plasmid vector).

The agarose gel was prepared in a 1% ratio (0.5g of agarose - 50 mL of TAE 1x buffer). TAE 1x was obtained by diluting 20 mL of TAE 50x in 980 mL of H2Obidest, as described below:

50x APR:

242 g Tris-base

57.1 ml acetic acid 100 mL EDTA 5.0M

842.9 mL $H_2O_{(bidest)}$ (to complete 1L)

When preparing the gel, while it was still in a liquid state, 2iil of SYBR Safe™ (Invitrogen) was added to enable the fragments to be visualised under UV light using a transilluminator.

Separation took place under an electric field of 50-70 V for 45 minutes.

CHAPTER 4

RESULTS AND DISCUSSION

The results of the PCR amplifications separated on agarose gels can be seen in figure 11:

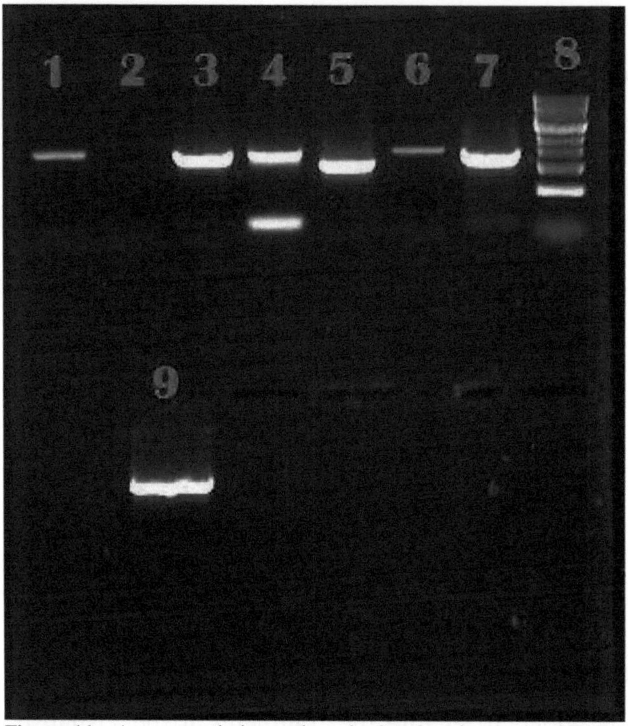

Figure 11 - Agarose gel electrophoresis showing the result of DNA amplification by PCR. Well 8 corresponds to the molecular weight marker, with fragment sizes in base pairs and in descending order equal to: 10,000 bp; 8,000 bp; 6,000 bp; 5,000 bp; 4,000 bp; 3,000 bp; 2,000 bp; 1,500 bp; 1,000 bp; 500 bp and 100 bp. The amplified sections appear clearly in wells 3, 4, 5 and 7, and weakly in sections 1 and 6. Nothing was amplified in well 2.

For a better understanding, table 5 describes the mixture of PCR components corresponding to each number illustrated in the figure:

Table 5 - Mixture of PCR components corresponding to each well, as shown in figure 11.

PCR	DNA	*Primers*	Expected size
1	DENV-2	DE2-F3R2	1488 pb

2	DENV-2	DE2-F4R3	1976 pb
3	DENV-4	DE4-F2R2	1486 pb
4	DENV-4	DE4-F3R3	1546 pb
5 e9	pGEX-2T	pGEX-F2R2	1146 pb
6	PCR 2	DE2-F3R2	1488 pb
7	PCR 4	DE4-F2R2	1486 pb
8	Marker	-	-

The expected sizes of the amplified fragments corresponded to:

- **DENV-2**: 1976 bp and 1488 bp using the external primers (DE2-F4R3) and internal primers (DE2-F3R2), respectively;
- **DENV-4**: 1546 and 1486 bp using the external primers (DE4-F3R3) and internal primers (DE4-F2R2), respectively;
- **pGEX-2T** vector: 1,146 bp for the amplification of the pGEX-2T vector fragment using the pGEX-F2R2 primers.

As can be seen in figure 11 corresponding to the agarose gel electrophoresis, the size of the fragments that were successfully amplified corresponds to the expected size of each section.

In **PCRs 6** and **7**, in which **PCR** products **2** and **4** were used, respectively, as the DNA template (Nested PCR) and the internal primers, fragments of sizes corresponding to the amplifications of PCRs 1 and 3 (DENV-2 and DENV-4 as the template, and internal primers) were produced in this order, as was also expected.

In **PCR 4** with the combination of the DENV-4 DNA molecule and external primers, two bands were observed, whereas only one would have been expected. This result indicates that at least one of the primers was mislabelled.

PCR 2 (Template: DENV-2, external primers) showed no amplification. **PCR 1** (Template: DENV-2, internal primers) and **PCR 6** (Template: PCR 2, internal primers) obtained amplification of the expected size, although they were more weakly amplified compared to the others.

This would indicate possible problems in the reaction mixture:

The result seen in **PCR 2** shows that no copies of the DENV-2 fragment were produced using the external primers selected. This can be corroborated by the fact that **PCR's 1** and **6** have exactly the same intensity, emphasising that only the internal primers, but not the external DENV-2 primers, worked.

A In addition to the problems mentioned with the *primers,* the poor amplification may also have been caused by the poor quality of the DENV-2 template.

Despite the results obtained for the serotype-2 DNA molecule, it is possible to use the amplification product with the internal *primers* in the insertion stage of the E gene into the cloning vector.

The amplification fragments corresponding to the DENV-4 DNA molecule, the DENV-4 PCR product and pGEX-2T could be strongly visualised, confirming that numerous copies of the DNA sought to be amplified in each reaction were obtained, as was the aim of this work. It can therefore be said that the PCR technique developed for these fragments was satisfactory.

CHAPTER 5

CONCLUSION

This work established techniques for amplifying the E gene of dengue viral serotypes 2 and 4, as well as for amplifying a fragment of the pGEX-2T cloning vector. Once determined, the techniques for amplifying the E gene are essential for obtaining recombinant DNA in the process of cloning and expressing the dengue E protein.

The PCR established for DENV-2, as well as for DENV-4, was more successful when using the combination of internal primers for amplification.

Other techniques need to be carried out in order to express the E protein, including the steps of binding the amplified E gene to the plasmid, transforming bacteria with the plasmid vector containing the recombinant DNA, and expressing and purifying the chimeric E protein resulting from the transgene, with Western blot control of the expression.

CHAPTER 6

BIBLIOGRAPHICAL REFERENCES

ALBERTS, B.; BRAY, D.; LEWIS, J.;RAFF,M.; ROBERTS, K. & WATSON J.D., **Molecular Biology of the Cell**. 5th Edition, Artmed, 2009.

ALCARAZ-ESTRADA, Sofia Lizeth; YOCUPICIO-MONROY, Martha; DEL ANGEL, Rosa María. Insights into dengue virus genome replication. **Future Virology**, v. 5, n. 5, p. 575-592, 2010.

ARAÚJO, Herbert Crisóstomo dos Santos. **Amplification of the pGEX-2t cloning vector in chemo-competent E. coli cells and digestion of the vector with the Sma I restriction enzyme**. 2015.

ARAÚJO, Josélio Maria Galvão de. **Dengue virus serotype 3 (DENV-3) in Brazil: studies on pathogenesis, replication sites, phylogeny and molecular evolution**. 2016. PhD Thesis.

AZEVEDO, Adriana de Souza et al. **Development of DNA vaccines against dengue virus based on the viral envelope protein**. 2011. PhD Thesis. Oswaldo Cruz Institute.

BARBOZA, Morgana Maria de Oliveira. **Genomic characterisation of dengue virus serotype 1 isolated in Ceará in 2011**. 2013.

BARROS, Lilian P. S. et al . Critical analysis of haematological and serological findings in patients with suspected Dengue. **Rev. Bras. Hematol. Hemoter.**, São Paulo, v. 30, n. 5, p. 363-366, Oct. 2008.

BARROS, Maria Creuza do Espírito Santo. **Expression of dengue virus proteins in insect cells using the Baculovirus Expression System.** 2007. 108 f. Dissertation (Master's Degree) - Molecular Pathology Course, University of Brasília, Brasília, 2007.

BRAZIL, Ministry of Health. Health Surveillance Secretariat. Department of Communicable Disease Surveillance. **Dengue: diagnosis and clinical management: adults and children.** 4. ed. - Brasília: Ministry of Health, 2013. 80 p.

BRICKS, Lucia Ferro. Vaccines for dengue: perspectives. **Paediatrics (São Paulo)**, v. 26, n. 4, p. 268-281, 2004.

BROWN, T. A. **Gene cloning and DNA analysis: an introduction**. Artmed 4ed, 240p, 2003.

CARDOSA, M. J. Dengue vaccine design: issues and challenges. **British Medical Bulletin**, v. 54, n. 2, p. 395-405, 1998.

CARUSO, Celia Sulzbacher. **Cloning, expression and characterisation of recombinant proteins from Xylella fastidiosa**. 2007. PhD Thesis. University of São Paulo.

CHAMBERS, Thomas J. et al. Flavivirus genome organisation, expression, and replication. **Annual Reviews in Microbiology**, v. 44, n. 1, p. 649-688, 1990.

CHAO, Luke H. et al. Sequential conformational rearrangements in flavivirus membrane fusion. **Elife**, v. 3, p. e04389, 2014.

CHIANG, Chen-Yi et al. Lipidated dengue-2 envelope protein domain III independently stimulates long-lasting neutralising antibodies and reduces the risk of antibody-dependent enhancement. **PLoS Negl Trop Dis**, v. 7, n. 9, p. e2432, 2013.

CLYDE, K.; KYLE, J. L.; HARRIS, E. Recent advances in deciphering viral and host determinants of dengue virus replication and pathogenesis. **Journal of Virology**, v. 80, n. 23, p. 11418-11431, 2006.

CORDEIRO, Marli Tenório. **Evolution of dengue in the state of Pernambuco, 1987-2006: epidemiology and molecular characterisation of circulating serotypes**. 2008. PhD Thesis.

CRILL, Wayne D.; CHANG, Gwong-Jen J. Localisation and characterisation of flavivirus envelope glycoprotein cross-reactive epitopes. **Journal of virology**, v. 78, n. 24, p. 13975- 13986, 2004.

FLORES, N. et al. Recovery of DNA from agarose gels stained with methylene blue. **Biotechniques**, n. 13, p. 203-205, 1992.

GUZMÁN, M. G. et al. Dengue: a continuing global threat. **Nature Reviews Microbiology**, London, p. S7-S16, 2010.

GUZMÁN, **M.** G.; **KOURÍ,** Gustavo. Dengue diagnosis, advances and challenges. **International journal of infectious diseases**, v. 8, n. 2, p. 69-80, 2004.

HENCHAL, E. A. & PUTNAK, J. R. The Dengue Viruses. **Clinical Microbiology Reviews** v. 3 (4), p. 376-396, 1990.

JAISWAL, S.; KHANNA, N.; SWAMINATHAN S. Replication-defective adenoviral vaccine vector for the induction of immune responses to Dengue virus type 2. **Journal Of Virology** v. 77 (23), p. 12907-12913, 2003.

KIM, Yong et al. The gene: the polymerase chain reaction and its clinical application. **Journal of oral and maxillofacial surgery**, v. 60, n. 7, p. 808-815, 2002.

KOSTYUCHENKO, V. A. et al. Near-atomic resolution cryo-electron microscopic structure of dengue serotype 4 virus. **Journal of Virology**, v. 88, n. 1, p. 477-482, 2014.

LEITMEYER, Katrin C. et al. Dengue virus structural differences that correlate with pathogenesis. **Journal of virology**, v. 73, n. 6, p. 4738-4747, 1999.

LIMA, L.M. **Basic Concepts of Molecular Biology Techniques.** Embrapa Cotton, Campina Grande, 2008.

LUDWIG, Jordana Beatris. **Cloning, expression and production of high fidelity**

dna polymerase from pyrococcus furiosus. 2010.

MACIEL, I. J.; SIQUEIRA JÚNIOR, J. B.; MARTELLI, C. M. T. **Epidemiology and challenges in dengue control.** Revista de Patologia Tropical, Goiânia, v. 37, n.2, 2008.

MONDINI, Adriano et al. Spatio-temporal tracking and phylodynamics of an urban dengue 3 outbreak in Sao Paulo, Brazil. **PLoS Negl Trop Dis**, v. 3, n. 5, p. e448, 2009.

MONGKOLSAPAYA, J. et al. Original antigenic sin and apoptosis in the pathogenesis of dengue haemorrhagic fever. **Nature medicine**, v. 9, n. 7, p. 921-927, 2003.

MUKHOPADHYAY, Suchetana; KUHN, Richard J.; ROSSMANN, Michael G. A structural perspective of the flavivirus life cycle. **Nature Reviews Microbiology**, v. 3, n. 1, p. 13-22, 2005.

MURRELL, Sarah; WU, Suh-Chin; BUTLER, Michael. Review of dengue virus and the development of a vaccine. **Biotechnology advances**, v. 29, n. 2, p. 239-247, 2011.

MUSTAFA, M. S. et al. Discovery of fifth serotype of dengue virus (DENV-5): A new public health dilemma in dengue control. **Medical Journal Armed Forces India,** v. 71, n. 1, p. 67- 70, 2015.

NASCIMENTO, Alexandra AC et al. **Recombinant DNA Technology.**

Ribeirão Preto: University of São Paulo, p. 85, 1999.

NELSON, D. L.; COX, M. M. **Lehninger's principles of biochemistry**. 6. ed. Porto Alegre: Artmed, 2014.

OLIVEIRA, André Silva de. **Heterologous expression of domain III of the envelope protein (E) of dengue virus 1 and 2 in** Pichia pastoris. 2011. 48 f. Dissertation (Master's) - Course in Cellular and Structural Biology, Universidade Federal de Viçosa, Minas Gerais, 2011.

OLIVEIRA, Anibal Silva de. **Cloning and expression of the NS1 and NS3 recombinant proteins of dengue virus type 3**. 2013. PhD Thesis. University of São Paulo.

PAN AMERICAN HEALTH ORGANISATION; PAN AMERICAN SANITARY BUREAU; PAN AMERICAN SANITARY BUREAU STAFF. **Dengue and dengue haemorrhagic fever in the Americas: guidelines for prevention and control**. Pan American Health Organisation, Pan American Sanitary Bureau, Regional Office of the World Health Organisation, 1995.

PAIXÃO, Vinicius Ferreira da. **Construction and heterologous expression of domain III of the envelope protein (E) of dengue-1e-3 viruses in Pichia pastoris**. 2014.

PINHO, Aryane Cruz Oliveira. **Diagnosis and molecular characterisation of the dengue virus circulating in the city of Salvador, Bahia, Brazil**. 2013.

POLONI, Telma Regina Ramos Silva. **Detection and typing of dengue virus by real-time RT-PCR.** 2009. 76 f. Dissertation (Master's) - Course in Biosciences Applied to Pharmacy, University of São Paulo - USP, Ribeirão Preto, 2009.

SAUTTO, Giuseppe et al. Possible future monoclonal antibody (mAb)-based therapy against arbovirus infections. **BioMed research international**, v. 2013, 2013.

SILVA, Ana Maria da. **Molecular characterisation of dengue viruses circulating in Pernambuco: epidemiological implications.** 2013. Thesis (Doctorate in Public Health) - Aggeu Magalhães Research Centre, Oswaldo Cruz Foundation, Recife, 2013.

SHRESTHA, Bimmi et al. The development of therapeutic antibodies that neutralise homologous and heterologous genotypes of dengue virus type 1.**PLoS Pathog**, v. 6, n. 4, p. e1000823, 2010.

SWISS INSTITUTE OF BIOINFORMATICS, Flaviviridae. Available at: <http://viralzone.expasy.org/viralzone/all_by_species/43.html>. Accessed on 02 September 2016.

UHLMANN, V. et al. In cell amplification. **Molecular Pathology**, v. 51, n. 3, p. 119, 1998.

WEAVER, R. F. **Molecular biology**. Lawrence: University of Kansas, 2001. 880 p.

WESTAWAY, E.G. et al. Flaviviridae. **Intervirology**, v. 24, n. 4, p. 183-92, 1985.

WHITEHEAD, Stephen S. et al. Prospects for a dengue virus vaccine.**Nature Reviews Microbiology**, v. 5, n. 7, p. 518-528, 2007.

YAP, Thai Leong et al. Crystal structure of the dengue virus RNA-dependent RNA polymerase catalytic domain at 1.85-angstrom resolution. **Journal of virology**, v. 81, n. 9, p. 4753-4765, 2007.

ZAHA, Arnaldo; FERREIRA, Henrique Bunselmeyer; PASSAGLIA, Luciane MP. **Basic Molecular Biology-5**. Artmed Editora, 2014.

ZHANG, Wei et al. Visualisation of membrane protein domains by cryo-electron microscopy of dengue virus. **Nature Structural & Molecular Biology**, v. 10, n. 11, p. 907-912, 2003.

CHAPTER 7

APPENDIX A

NUCLEOTYPICAL SEQUENCE OF THE pGEX-2T VECTOR, AMPLIFIED FRAGMENT AND LOCALISATION OF PRIMERS

Complete nucleotide sequence of the circular genome of the pGEX-2T vector, totalling 4948 bp: o

- The amplified fragment of the vector (highlighted in **yellow**) contains 1146 bp and the respective primers used for this amplification are **underlined and in bold**.

ACGTTATCGACTGCACGGTGCACCAATGCTTCTGGCGTCAGGCAGCCATC
GGAAGCTGTGGTATGGCTGTGCAGGTCGTAAATCACTGCATAATTCGTGT
CGCTCAAGGCGCACTCCCGTTCTGGATAATGTTTTTTGCGCCGACATCATA
ACGGTTCTGGCAAATATTCTGAAATGAGCTGTTGACAATTAATCATCGGC
TCGTATAATGTGTGGAATTGTGAGCGGATAACAATTTCACACAGGAAACA
GTATTCATGTCCCCTATACTAGGTTATTGGAAAATTAAGGGCCTTGTGCA
ACCCACTCGACTTCTTTTGGAATATCTTGAAGAAAAATATGAAGAGCATT
TGTATGAGCGCGATGAAGGTGATAAATGGCGAAACAAAAAGTTTGAATT
GGGTTTGGAGTTTCCCAATCTTCCTTATTATATTGATGGTGATGTTAAATT
AACACAGTCTATGGCCATCATACGTTATATAGCTGACAAGCACAACATGT
TGGGTGGTTGTCCAAAAGAGCGTGCAGAGATTTCAATGCTTGAAGGAGCG
GTTTTGGATATTAGATACGGTGTTTCGAGAATTGCATATAGTAAAGACTTT
GAAACTCTCAAAGTTGATTTTCTTAGCAAGCTACCTGAAATGCTGAAAAT
GTTCGAAGATCGTTTATGTCATAAAACATATTTAAATGGTGATCATGTAA
CCCATCCTGACTTCATGTTGTATGACGCTCTTGATGTTGTTTTATACATGG
ACCCAATGTGCCTGGATGCGTTCCCAAAATTAGTTTGTTTTAAAAAACGT
ATTGAAGCTATCCCACAAATTGATAAGTACTTGAAATCCAGCAAGTATAT
AGCATGGCCTTTGCAGGGCTGGCAAGCCACGTTTGGTGGTGGCGACCATC
CTCCAAAATCGGATCTGGTTCCGCGTGGATCCCCGGGAATTCATCGTGAC
TGACTGACGATCTGCCTCGCGCGTTTCGGTGATGACGGTGAAAACCTCTG
ACACATGCAGCTCCCGGAGACGGTCACAGCTTGTCTGTAAGCGGATGCCG
GGAGCAGACAAGCCCGTCAGGGCGCGTCAGCGGGTGTTGGCGGGTGTCG
GGGCGCAGCCATGACCCAGTCACGTAGCGATAGCGGAGTGTATAATTCTT
GAAGACGAAAGGGCCTCGTGATACGCCTATTTTTATAGGTTAATGTCATG
ATAATAATGGTTTCTTAGACG**TCAGGTGGCACTTTTCGGGG**AAATGTGC
GCGGAACCCCTATTTGTTTATTTTTCTAAATACATTCAAATATGTATCCGC
TCATGAGACAATAACCCTGATAAATGCTTCAATAATATTGAAAAAGGAAG
AGTATGAGTATTCAACATTTCCGTGTCGCCCTTATTCCCTTTTTTGCGGCA
TTTTGCCTTCCTGTTTTTGCTCACCCAGAAACGCTGGTGAAAGTAAAAGAT

GCTGAAGATCAGTTGGGTGCACGAGTGGGTTACATCGAACTGGATCTCAA
CAGCGGTAAGATCCTTGAGAGTTTTCGCCCCGAAGAACGTTTTCCAATGA
TGAGCACTTTTAAAGTTCTGCTATGTGGCGCGGTATTATCCCGTGTTGACG
CCGGGCAAGAGCAACTCGGTCGCCGCATACACTATTCTCAGAATGACTTG
GTTGAGTACTCACCAGTCACAGAAAAGCATCTTACGGATGGCATGACAGT
AAGAGAATTATGCAGTGCTGCCATAACCATGAGTGATAACACTGCGGCCA
ACTTACTTCTGACAACGATCGGAGGACCGAAGGAGCTAACCGCTTTTTTG
CACAACATGGGGGATCATGTAACTCGCCTTGATCGTTGGGAACCGGAGCT
GAATGAAGCCATACCAAACGACGAGCGTGACACCACGATGCCTGCAGCA
ATGGCAACAACGTTGCGCAAACTATTAACTGGCGAACTACTTACTCTAGC
TTCCCGGCAACAATTAATAGACTGGATGGAGGCGGATAAAGTTGCAGGA
CCACTTCTGCGCTCGGCCCTTCCGGCTGGCTGGTTTATTGCTGATAAATCT
GGAGCCGGTGAGCGTGGGTCTCGCGGTATCATTGCAGCACTGGGGCCAG
ATGGTAAGCCCTCCCGTATCGTAGTTATCTACACGACGGGGAGTCAGGCA
ACTATGGATGAACGAAATAGACAGATCGCTGAGATAGGTGCCTCACTGAT
TAAGCATTGGTAACTGTCAGACCAAGTTTACTCATATATACTTTAGATTGA
TTTAAAACTTCATTTTTAATTTAAAAGGATCTAGGTGAAGATCCTTTTTGA
TAATCTCATGACCAAAATCCCTTAACGTGAGTTTTCGTTCCACT**GAGCGT**
CAGACCCCGTAGAAAAGATCAAAGGATCTTCTTGAGATCCTTTTTTTCTG
CGCGTAATCTGCTGCTTGCAAACAAAAAAACCACCGCTACCAGCGGTGGT
TTGTTTGCCGGATCAAGAGCTACCAACTCTTTTTCCGAAGGTAACTGGCTT
CAGCAGAGCGCAGATACCAAATACTGTCCTTCTAGTGTAGCCGTAGTTAG
GCCACCACTTCAAGAACTCTGTAGCACCGCCTACATACCTCGCTCTGCTA
ATCCTGTTACCAGTGGCTGCTGCCAGTGGCGATAAGTCGTGTCTTACCGG
GTTGGACTCAAGACGATAGTTACCGGATAAGGCGCAGCGGTCGGGCTGA
ACGGGGGGTTCGTGCACACAGCCCAGCTTGGAGCGAACGACCTACACC
GAACTGAGATACCTACAGCGTGAGCTATGAGAAAGCGCCACGCTTCCCG
AAGGGAGAAAGGCGGACAGGTATCCGGTAAGCGGCAGGGTCGGAACAG
GAGAGCGCACGAGGGAGCTTCCAGGGGGAAACGCCTGGTATCTTTATAG
TCCTGTCGGGTTTCGCCACCTCTGACTTGAGCGTCGATTTTTGTGATGCTC
GTCAGGGGGGCGGAGCCTATGGAAAAACGCCAGCAACGCGGCCTTTTTA
CGGTTCCTGGCCTTTTGCTGGCCTTTTGCTCACATGTTCTTTCCTGCGTTAT
CCCCTGATTCTGTGGATAACCGTATTACCGCCTTTGAGTGAGCTGATACCG
CTCGCCGCAGCCGAACGACCGAGCGCAGCGAGTCAGTGAGCGAGGAAGC
GGAAGAGCGCCTGATGCGGTATTTTCTCCTTACGCATCTGTGCGGTATTTC
ACACCGCATAAATTCCGACACCATCGAATGGTGCAAAACCTTTCGCGGTA
TGGCATGATAGCGCCCGGAAGAGAGTCAATTCAGGGTGGTGAATGTGAA
ACCAGTAACGTTATACGATGTCGCAGAGTATGCCGGTGTCTCTTATCAGA
CCGTTTCCCGCGTGGTGAACCAGGCCAGCCACGTTTCTGCGAAAACGCGG
GAAAAAGTGGAAGCGGCGATGGCGGAGCTGAATTACATTCCCAACCGCG
TGGCACAACAACTGGCGGGCAAACAGTCGTTGCTGATTGGCGTTGCCACC
TCCAGTCTGGCCCTGCACGCGCCGTCGCAAATTGTCGCGGCGATTAAATC
TCGCGCCGATCAACTGGGTGCCAGCGTGGTGGTGTCGATGGTAGAACGAA
GCGGCGTCGAAGCCTGTAAAGCGGCGGTGCACAATCTTCTCGCGCAACGC

GTCAGTGGGCTGATCATTAACTATCCGCTGGATGACCAGGATGCCATTGC
TGTGGAAGCTGCCTGCACTAATGTTCCGGCGTTATTTCTTGATGTCTCTGA
CCAGACACCCATCAACAGTATTATTTTCTCCCATGAAGACGGTACGCGAC
TGGGCGTGGAGCATCTGGTCGCATTGGGTCACCAGCAAATCGCGCTGTTA
GCGGGCCCATTAAGTTCTGTCTCGGCGCGTCTGCGTCTGGCTGGCTGGCA
TAAATATCTCACTCGCAATCAAATTCAGCCGATAGCGGAACGGGAAGGC
GACTGGAGTGCCATGTCCGGTTTTCAACAAACCATGCAAATGCTGAATGA
GGGCATCGTTCCCACTGCGATGCTGGTTGCCAACGATCAGATGGCGCTGG
GCGCAATGCGCGCCATTACCGAGTCCGGGCTGCGCGTTGGTGCGGATATC
TCGGTAGTGGGATACGACGATACCGAAGACAGCTCATGTTATATCCCGCC
GTTAACCACCATCAAACAGGATTTTCGCCTGCTGGGGCAAACCAGCGTGG
ACCGCTTGCTGCAACTCTCTCAGGGCCAGGCGGTGAAGGGCAATCAGCTG
TTGCCCGTCTCACTGGTGAAAAGAAAAACCACCCTGGCGCCCAATACGCA
AACCGCCTCTCCCCGCGCGTTGGCCGATTCATTAATGCAGCTGGCACGAC
AGGTTTCCCGACTGGAAAGCGGGCAGTGAGCGCAACGCAATTAATGTGA
GTTAGCTCACTCATTAGGCACCCCAGGCTTTACACTTTATGCTTCCGGCTC
GTATGTTGTGTGGAATTGTGAGCGGATAACAATTTCACACAGGAAACAGC
TATGACCATGATTACGGATTCACTGGCCGTCGTTTTACAACGTCGTGACTG
GGAAAACCCTGGCGTTACCCAACTTAATCGCCTTGCAGCACATCCCCCTT
TCGCCAGCTGGCGTAATAGCGAAGAGGCCCGCACCGATCGCCCTTCCCAA
CAGTTGCGCAGCCTGAATGGCGAATGGCGCTTTGCCTGGTTTCCGGCACC
AGAAGCGGTGCCGGAAAGCTGGCTGGAGTGCGATCTTCCTGAGGCCGAT
ACTGTCGTCGTCCCCTCAAACTGGCAGATGCACGGTTACGATGCGCCCAT
CTACACCAACGTAACCTATCCCATTACGGTCAATCCGCCGTTTGTTCCCAC
GGAGAATCCGACGGGTTGTTACTCGCTCACATTTAATGTTGATGAAAGCT
GGCTACAGGAAGGCCAGACGCGAATTATTTTTGATGGCGTTGGAATT

(References: pGEX-2T cloning vector, complete sequence; GenBank: U13850.1)

APPENDIX B

DENV-2 NUCLEOTIDE SEQUENCE AND LOCATION OF PRIMERS **IN THE VIRAL GENOME**

Part of the nucleotide sequence of the DENV-2 gene, which includes the E gene:

- prM gene (**yellow**), where the external forward primer starts synthesis;

- Gene E (**blue**), where the forward and reverse internal primers start and finish synthesis;

- Part of the genome encoding the NS1 non-structural protein (**grey**): this is where the external reverse primer used to amplify the E gene is located.

- The external primers (**underlined and in bold**), and the internal primers (**bold**).

TTCCATTTAACCACACGTAATGGAGAACCACACATGATC**GTGGGGAGGC**
AAGAGAAAGGGAAAAGTCTTCTGTTCAAAACAGAGGATGGTGTCAACAT
GTGCACCCTCATGGCCATAGACCTTGGTGAATTGTGTGAAGATACAATCA
CGTACAAGTGTCCCCTCCTCAGGCAAAATGAACCAGAAGACATAGATTGT
TGGTGCAACTCTACGTCCACATGGGTAACTTATGGGACATGTACCACCAC
AGGAGAACACAGAAGAGAAAAAAGATCAGTGGCACTCGTTCCACATGTG
GGTATGGGACTGGAGACACGAACTGAAACATGGATGTCATCAGAAGGG
GCCTGGAAACATGTTCAGAGAATTGAAACCTGGATCTTGAGACATCCAGG
CTTTGCCATAATGGCAGCAATCCTGGCATACACCATAGGAACGACACATT
TCCAAAGGGCCTTGATTTTCATCTTACTGACAGCTGTCGCTCCTTCAATGA
C**AATGCGTTGCATAGGAATTTCAAAT**AGAGACTTCGTAGAAGGGGTTTC
AGGAGGAAGCTGGGTTGACATAGTCTTAGAACATGGAAGTTGTGTGACG
ACGATGGCAAAAAATAAACCAACATTGGATTTTGAACTGATAAAAACAG
AAGCCAAACAACCTGCCACTCTAAGGAAGTACTGTATAGAAGCAAAGCT
GACCAATACAACAACAGAATCTCGTTGCCCAACACAAGGGGAACCCAGT
CTAAATGAAGAGCAGGACAAAAGGTTCATCTGCAAACACTCCATGGTAG
ACAGAGGATGGGGAAATGGATGTGGATTATTTGGAAAAGGAGGCATTGT
GACCTGTGCTATGTTCACATGCAAAAAGAACATGGAAGGAAAAGTCGTG
CTGCCAGAAAATCTGGAATACACCATCGTGATAACACCTCACTCAGGAGA
AGAGCACGCTGTAGGTAATGACACAGGAAAACATGGCGAGGAAATCAAA
ATAACACCACAGAGTTCCATCACAGAAGCAGAACTGACAGGCTATGGCA
CTGTCACGATGGAGTGCTCTCCGAGAACGGGCCTCGACTTCAATGAAATG
GTGCTGCTGCAGATGGAAGAGAAAGCTTGGCTGGTGCACAGGCAATGGT
TCCTAGACCTGCCATTACCATGGCTACCCGGAGCGGACACACAAGGATCA
AATTGGATACAGAAAGAGACATTGGTCACTTTCAAAAATCCCCACGCGAA

GAAACAAGATGTCGTTGTTTTAGGATCTCAAGAAGGGGCCATGCACACGG
CACTCACAGGGGCCCAAAAAATCCAGTTGGCAATCAGAAACTTACTGTTC
ACAGGACATCTCAAGTGCAGGCTGAGAATGGACAAACTACAGCTCAAAG
GAATGTCATACTCTATGTGTACAGGAAAGTTTAAAATTGTGAAGGAAATA
GCAGAAACACAACATGGAACAATAGTTATCAGAGTACAATATGAAGGGG
ACGGCTCTCCATGTAAGATCCCTTTTGAGATAATGGATTTGGAAAAAAGA
CACGTCTTAGGTCGCCTGATTACAGTTAACCCAATCGTAACAGAAAAAGA
TAGCCCAGTCAACATAGAAGCAGAACCTCCATTCGGAGACAGCTACATCA
TCATAGGAGTAGAGCCGGGACAATTGAAACTCAATTGGTTTAAGAAGGG
AAGTTCCATCGGCCAAATGTTTGAGACAACAATGAGAGGAGCGAAGAGA
ATGGCCATATTAGGTGACACAGCCTGGGATTTTGGATCCCTGGGAGGAGT
GTTTACATCTATAGGAAAGGCTCTCCACCAAGTTTTCGGAGCAATCTATG
GGGCTGCTTTTAGTGGGGTCTCATGGACTATGAAAATCCTCATAGGAGTC
ATCATCACATGGATAGGAATGAATTCACGTAGCACCTCACTGTCTGTGTC
ACTAGTATTGGTGGGAGTCGTGACACTGTACCTGG**GAGCTATGGTGCAG**
GCTGATCAAGGATG**CGCCATCAACTTTGGCAAGAG**AGAGCTCAAGTGC
GGAGATGGTATCTTCATATTTAGAGACTCTGATGACTGGCTGAACAAGTA

(References: Dengue virus 2 isolate 57135/BR-PE/99, complete genome; GenBank: JX669486.1)

APPENDIX C

DENV-4 NUCLEOTIDE SEQUENCE AND LOCATION OF PRIMERS IN THE VIRAL GENOME

Part of the nucleotide sequence of the DENV-4 gene, which includes the E gene:

- prM gene (**yellow**), where the external forward primer starts synthesis;

- Gene E (**blue**), where the forward and reverse internal primers start and finish synthesis;

- Part of the genome encoding the NS1 non-structural protein (**grey**): this is where the external reverse primer used to amplify the E gene is located.

- The external primers (**underlined and in bold**), and the internal primers (**bold**).

AGACCTCTCTTGTTTAAGACAACAGAGGGGATCAACAAATGCACTCTCA
TTGCCATGGACTTGGGTGAAATGTGTGAGGACACTGTCACGTATAAATG
CCCCCTACTGGTCAATACCGAACCTGAAGACATTGATTGCTGGTGCAAC
CTCACGTCTACCTGGGTCATGTATGGGACATGCACCCAGAGCGGAGAAC
GGAGACGAGAGAAGCGCTCAGTAGCTTTAACACCACATTCAGGAATGG
GATTGGAAACAAGAGCTGAGACATGGATGTCATCGGAAGGGGCTTGGA
AGCATGCTCAGAGAGTAGAGAGCTGGATACTCAGAAACCCAGGATTCG
CGCTCTTGGCAGGATTTATGGCTTATATGATTGGGCAAACAGGAATCCA
GCGAACTGTCTTCTTCGT**CCTAATGATGCTGGTCGCCCC**ATCCTACGG
AATGCGATGCGTAGGAGTAGGAAACAGAGACTTTGTGGAAGGAGTCT
CAGGTGGAGCATGGGTCGACCTGGTGCTAGAACATGGAGGATGCGTCA
CAACCATGGCCCAGGGAAAACCAACCTTGGATTTTGAACTGACTAAGA
CAACAGCCAAGGAAGTGGCTCTGTTAAGAACCTATTGCATTGAAGCCTC
AATATCAAACATAACTACGGCAACAAGATGTCCAACGCAAGGAGAGCC
TTATCTGAAAGAGGAACAGGACCAACAGTACATTTGCCGGAGAGATGT
GGTAGACAGAGGGTGGGGCAATGGCTGTGGCTTGTTTGGAAAAGGAGG
AGTTGTGACATGTGCTAAGTTTTCATGTTCGGGGAAGATAACAGGCAAT
TTGGTCCAAATTGAGAACCTTGAATACACAGTGGTTGTAACAGTCCACA
ATGGAGACACCCATGCAGTAGGAAATGACACATCCAATCATGGAGTTA
CAGCCATGATAACTCCCAGGTCACCATCGGTGGAAGTCAAATTTCCGGA
CTATGGAGAACTAACACTCGATTGTGAACCCAGGTCTGGAATTGACTTT
AATGAGATGATTCTGATGAAAATGAAAAAGAAAACATGGCTCGTGCAT
AAGCAATGGTTTTTGGATCTGCCTCTTCCATGGACAGCAGGAGCAGACA
CATCAGAGGTTCACTGGAATTACAAAGAGAGAATGGTGACATTTAAGG
TTCCTCATGCCAAGAGACAGGATGTGACAGTGCTGGGATCTCAGGAAG

GAGCCATGCATTCTGCCCTCGCTGGAGCCACAGAAGTGGACTCCGGTGA
TGGAAATCACATGTTTGCAGGACATCTCAAGTGCAAAGTCCGTATGGAG
AGATTGAGAATCAAGGGAATGTCATACACGATGTGTTCAGGAAAGTTTT
CAATTGACAAAGAGATGGCAGAAACACAGCATGGGACAACAGTGGTGA
AAGTCAAGTATGAAGGTGCTGGAGCTCCGTGTAAAGTCCCCATAGAGA
TAAGAGATGTAAACAAGGAAAAAGTGGTTGGGCGTATCATCTCATCCA
CCCCTTTGGCTGAGAATACCAACAGTGTAACCAACATAGAATTAGAACC
CCCCTTTGGGGACAGCTACATAGTGATAGGTGTTGGAAACAGCGCATTA
ACACTCCACTGGTTCAGGAAAGGGAGTTCCATTGGCAAGATGTTTGAGT
CCACATACAGAGGTGCAAAACGAATGGCCATTCTAGGTGAAACAGCTT
GGGATTTTGGTTCCGTTGGTGGACTGTTCACATCATTGGGAAAGGCTGT
GCACCAGGTTTTTGGAAGTGTGTACACAACCATGTTTGGAGGAGTCTCA
TGGATGATTAGAATCCTAATTGGGTTCTTAGTGTTGTGGATTGGCACGA
ACTCAAGGAACACTTCAATGGCTATGACGTGCATAGCTGTTGGAGGAAT
CACTCTGTTTCT**GGGCTTCACAGTTGGCGCC**GATCAA**GGATGCGCCA**
TCAACTTTGGCAAGAGAGAGCTCAAG

(References: Dengue virus 4 isolate BR/SJRP/850/2013, complete genome; GenBank: KP188566.1)

MIX
Papier aus verantwortungsvollen Quellen
Paper from responsible sources
FSC® C105338

Printed by Books on Demand GmbH, Norderstedt / Germany